Progress with Oxford

Age 3-4

Numbers up to 10

Hello! I'm Dijit and this is Fijit.

Contents

Key

 Count

 Write

 Say

 Colour

 Draw

 Match

 Circle

 Trace with finger

 Trace with pencil

 Play together

 Find the sticker

OXFORD
UNIVERSITY PRESS

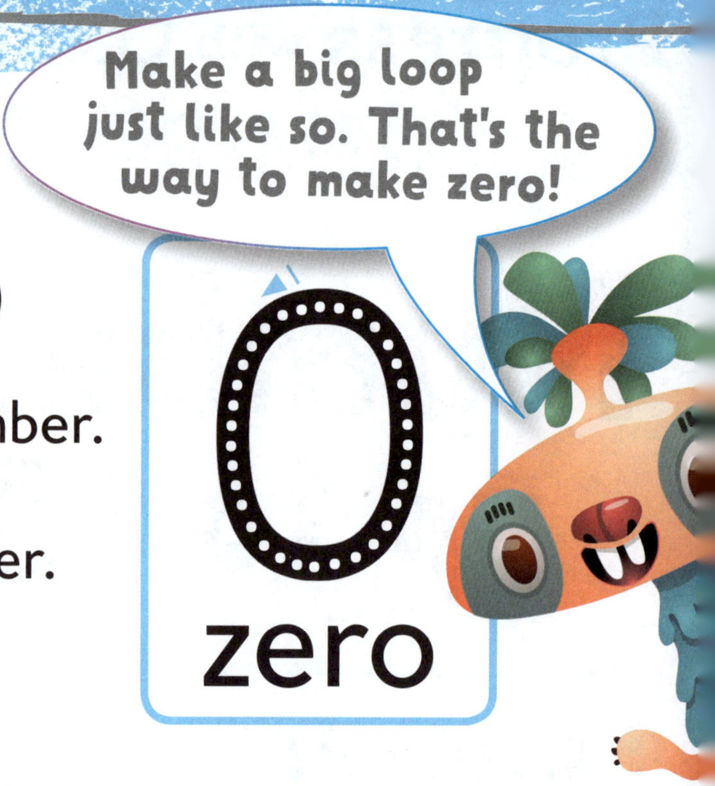

Make a big loop just like so. That's the way to make zero!

Number 0

 Trace the number.

 Say the number.

0 zero

 Trace these numbers.

 Circle every zero.

4 0 0 1

2 3 1

0 4 0 5

 ★★ Circle Fijit with **0** balloons.

 Write the number.

 Where can you find the number **0** in your house?

Well done!

 Play with numbers.

Use a tray of sand, flour or similar to practise writing **0**.

Be a number detective! Find number **0** in books at home.

Use play dough to make the number **0**.

Give yourself a sticker

Now – track how you're doing on page 32!

Number 1

 Trace the number.

 Say the number.

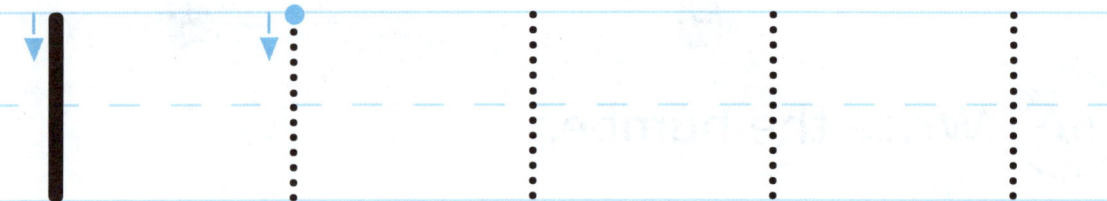

one

A straight line down and then you're done – that's how you make a number one.

 Trace these numbers.

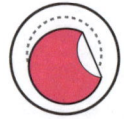 Find the sticker for each tool.

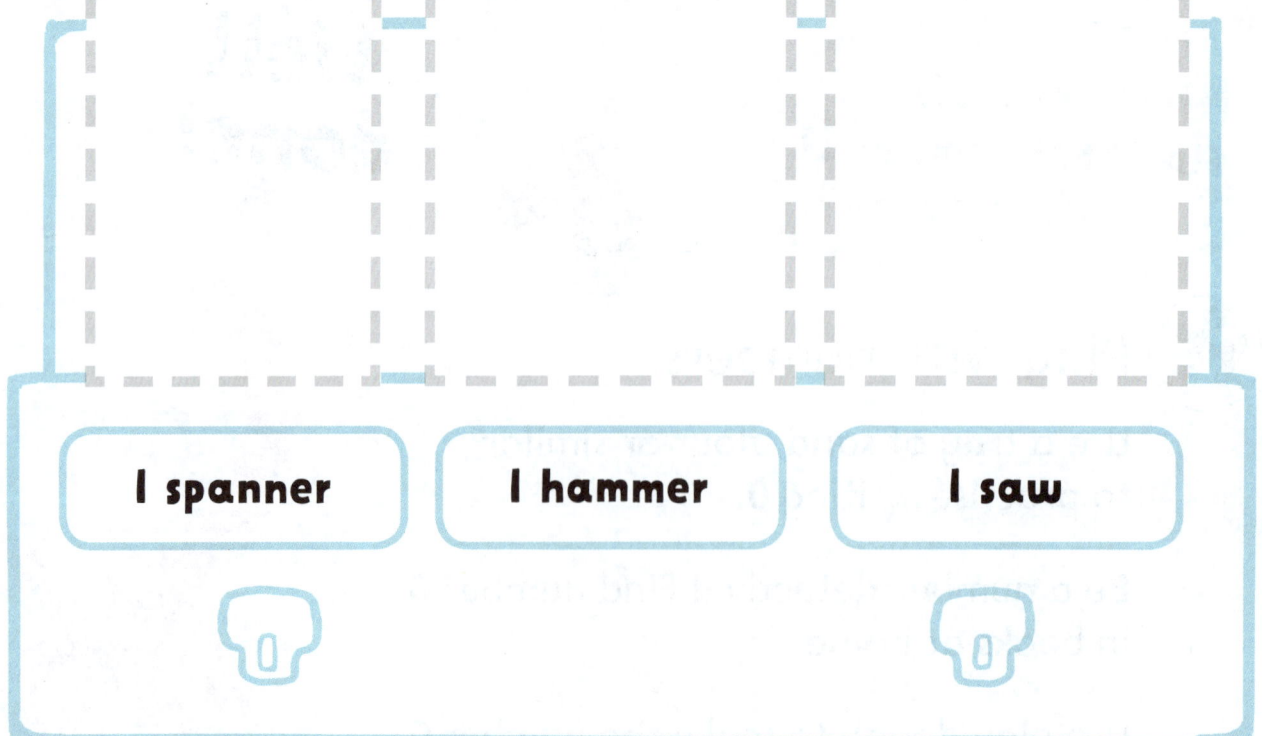

1 spanner

1 hammer

1 saw

 **Circle groups of 1.**

 **Write the number.**

Have more fun with the number 1.

Well done!

 **Play with numbers.**

Be a number detective! How many times can you find the number 1 on a walk to the shops?

Draw one boat with one sail and write a number 1 on the sail.

Write the number 1 on the path or an outside wall using a paintbrush dipped in water.

Give yourself a sticker

Now – track how you're doing on page 32!

Number 2

 Trace the number.

 Say the number.

2
two

Around and back
on the railroad track.
2, 2, 2, 2!

 Trace these numbers.

2 2 2 2 2

★★ Circle dice showing **2**.

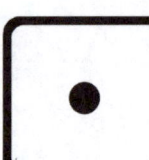
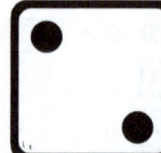

 Draw a line from groups of two objects to **2**.

 Write the number.

♪ **What number songs do YOU know?** ♪

 Play with numbers.

Sing *If you're happy and you know it* or another action song and do each action twice.

A pair means two of something that go together. How many different pairs of things can you find in your home?

Go for a walk. Collect two sticks, two leaves and two stones.

Well done!

Give yourself a sticker

Now – track how you're doing on page 32!

Around a tree, around a tree, that's the way to make a **3**.

Number 3

 Trace the number.

 Say the number.

3
three

 Trace these numbers.

3 3 3 3 3

 Colour **3** sheep, **3** cows and **3** horses.

 ★★ **Circle groups of 3 sheep.**

 **Write the number.**

3

 **Play with numbers.**

Have a picnic with 3 cuddly toys. Feed them three cakes, three biscuits or three apples.

Read *Goldilocks and the Three Bears* or the *Three Billy Goats Gruff* story with a grown-up.

Draw your favourite animal three times.

Well done!

Give yourself a sticker

Can you think of other stories with number 3 in the title?

Now – track how you're doing on page 32!

Number 4

 Trace the number.

 Say the number.

four

 Trace these numbers.

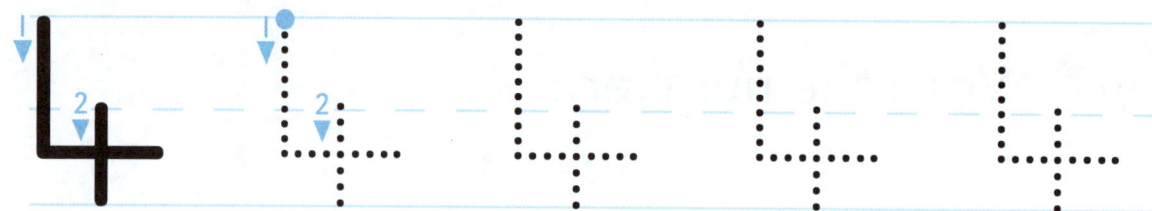

 Colour groups of 4.

⬤ Stick the number **4** stickers on the number 4s.

4 5 4 3

1 2 4

4 3 5

Can you find them all?

✏️ Write the number.

4

🤸 Play with numbers.

Find four different coloured toys.

On a piece of paper, practise writing the number 4. Draw four things under the number you wrote, or a monster with four arms, four eyes and four ears.

Make a slice of toast and talk about cutting it into two and then four.

Give yourself a sticker

Now – track how you're doing on page 32!

> Straight neck, big round tummy, hat on top, **5** looks funny!

Number 5

 Trace the number.

 Say the number.

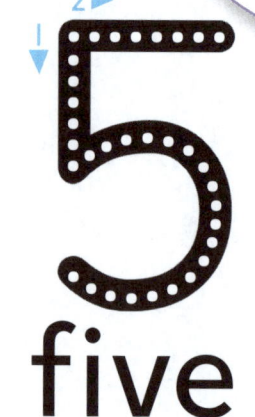
5
five

 Trace these numbers.

 Colour all the ways of showing **5**.

zero

five

4

four

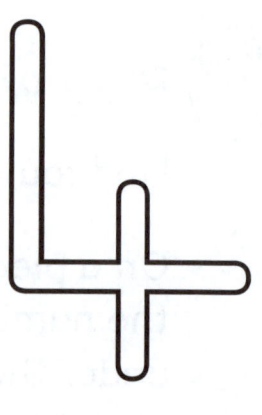

 Draw a line from groups of five to the number **5**.

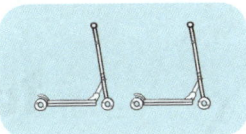

 Write the number.

 Play with numbers.

How many fingers do you have on one hand?

On a walk or a car journey count five cars, five lorries and five bikes.

Count your fingers and share a high five!

With a partner, take turns to challenge each other to find five things like five toy cars or five blue things.

Give yourself a sticker

Now – track how you're doing on page 32!

Numbers 0–5

Match the objects and words to the number.

Practise tracing the numbers as well!

0

1

2

3

4

5

five

two

one

four

zero

three

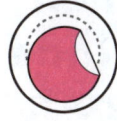 Find the sticker of the ladybird with spots that match the number on the leaf.

Count the spots before you add the sticker.

2

1

3

5

 Colour the bee using the right colour for each part.

0 1 2 3 4 5

Give yourself a sticker

Now – track how you're doing on page 32!

Number 6

 Trace the number.

 Say the number.

6
six

Make a curve, then make a loop. There are no tricks to make a six.

 Trace these numbers.

6 6 6 6 6

 Match the **6** caterpillars to **6** butterflies.

Stickers for page 4

Stickers for page 11

Stickers for page 15

Stickers for page 19

Reward Stickers

Match bugs with 6 legs to the number 6.

Practise counting to 6!

6

Write the number.

6

Play with numbers.

Count six flowers, six leaves and six minibeasts in the garden or park.

Complete a gym session – do six star jumps, six hops and take six giant steps.

Count how many eggs are in an egg box.

Give yourself a sticker

Now – track how you're doing on page 32!

Number 7

 Trace the number.

 Say the number.

7
seven

Right across, that's a given, then all the way down to make a seven.

 Trace these numbers.

7 7 7 7 7

 ★★ Circle groups of **7**.

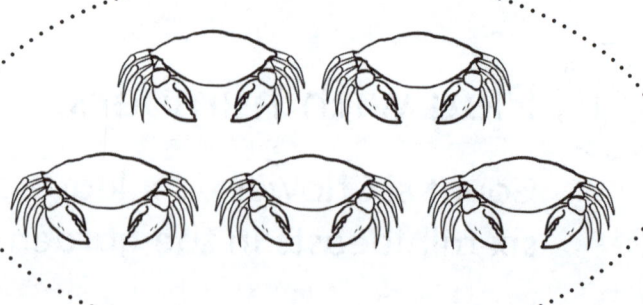

 Find **7** bucket stickers to match the 7 spades.

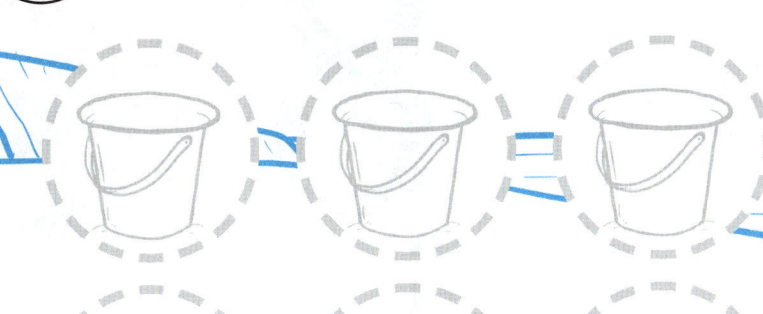

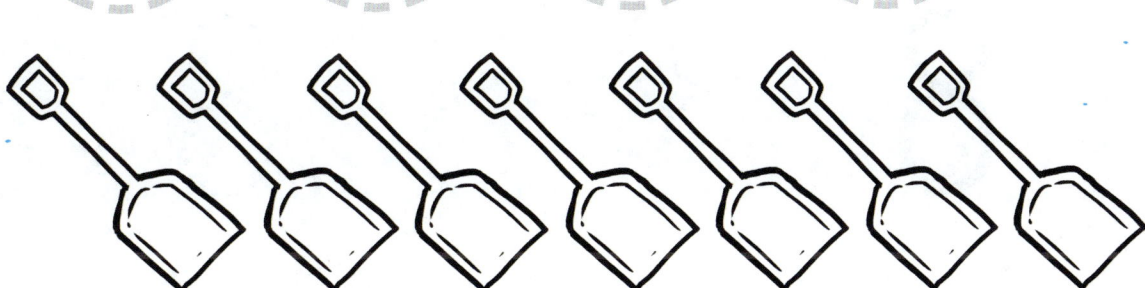

 Write the number.

7

Where can you find the number 7?

 Play with numbers.

Draw seven beach balls and colour them in.

Find seven stones or leaves in your garden or the park.

Build a tower using seven blocks.

Give yourself a sticker

Now – track how you're doing on page 32!

Number 8

 Trace the number.

 Say the number.

8
eight

 Trace these numbers.

 Draw a line from groups of **8** fruit to the bowl.

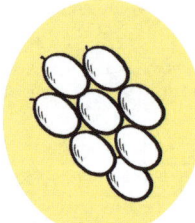  

 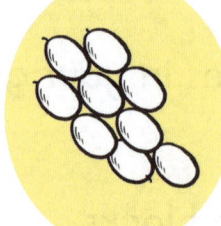 

 Colour **8** of each fruit or vegetable.

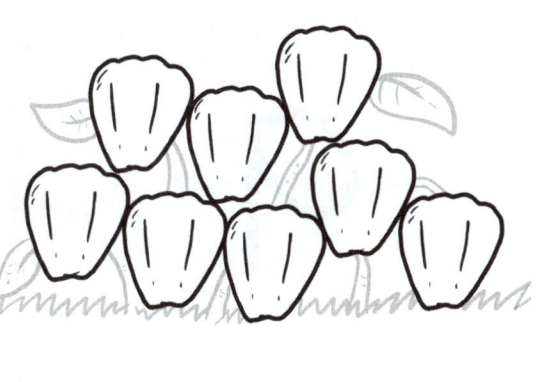

 Write the number.

8

 Play with numbers.

Count out eight toys, e.g. cars, marbles, crayons.

Use a scarf, ribbon or your finger to draw a large number 8 in the air.

Make the number 8 from modelling clay or pastry.

Give yourself a sticker

Now – track how you're doing on page 32!

Number 9

 Trace the number.

 Say the number.

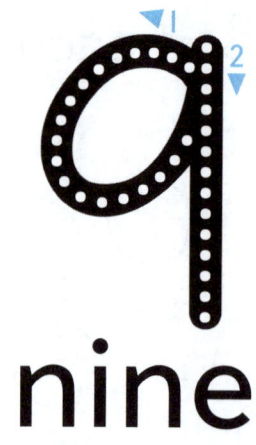
nine

A loop and a line, that makes 9!

 Trace these numbers.

 Colour all nines.

3	9	9	9	4
1	9	6	9	2
7	9	9	9	8
2	0	5	9	1
8	6	1	9	0
4	3	7	9	5

What can you see?

What can you do **9** times?

 Match groups of nine objects to **9**.

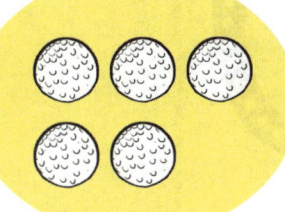

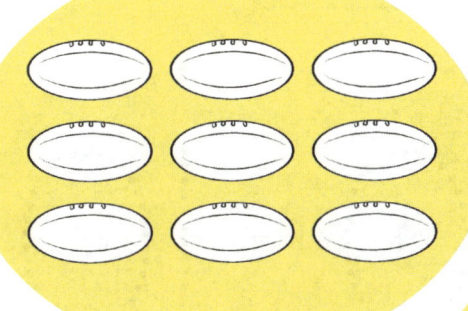

 **9**

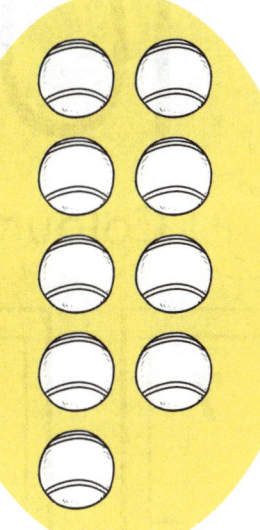

 Write the number.

 Play with numbers.

Bounce or hit a ball nine times.

Be a number detective! Find the number 9 in nine places in your home?

Give yourself a sticker

Now – track how you're doing on page 32!

Number 10

 Trace the number.

 Say the number.

10
ten

 Trace these numbers.

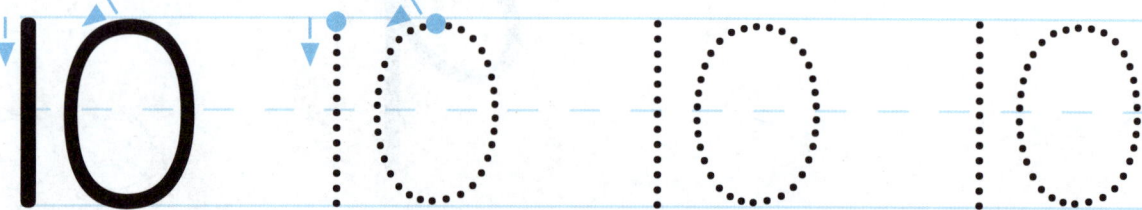

 Colour the front doors showing **10**.

Knock, knock!

Who's there?

 Circle all the **10**s you can find in this picture.

10 green bottles

 Write the number.

10

 Play with numbers.

Sing the song *10 green bottles* or *10 men went to mow*.

Line up ten toy bricks or other small objects.

Draw round both of your hands and then write the numbers 1, 2, 3, 4, 5, 6, 7, 8, 9 and 10 in order next to your fingers and thumbs.

Give yourself a sticker

Now – track how you're doing on page 32!

Numbers 6–10

 Match the objects and words to the number.

6

7

8

9

10

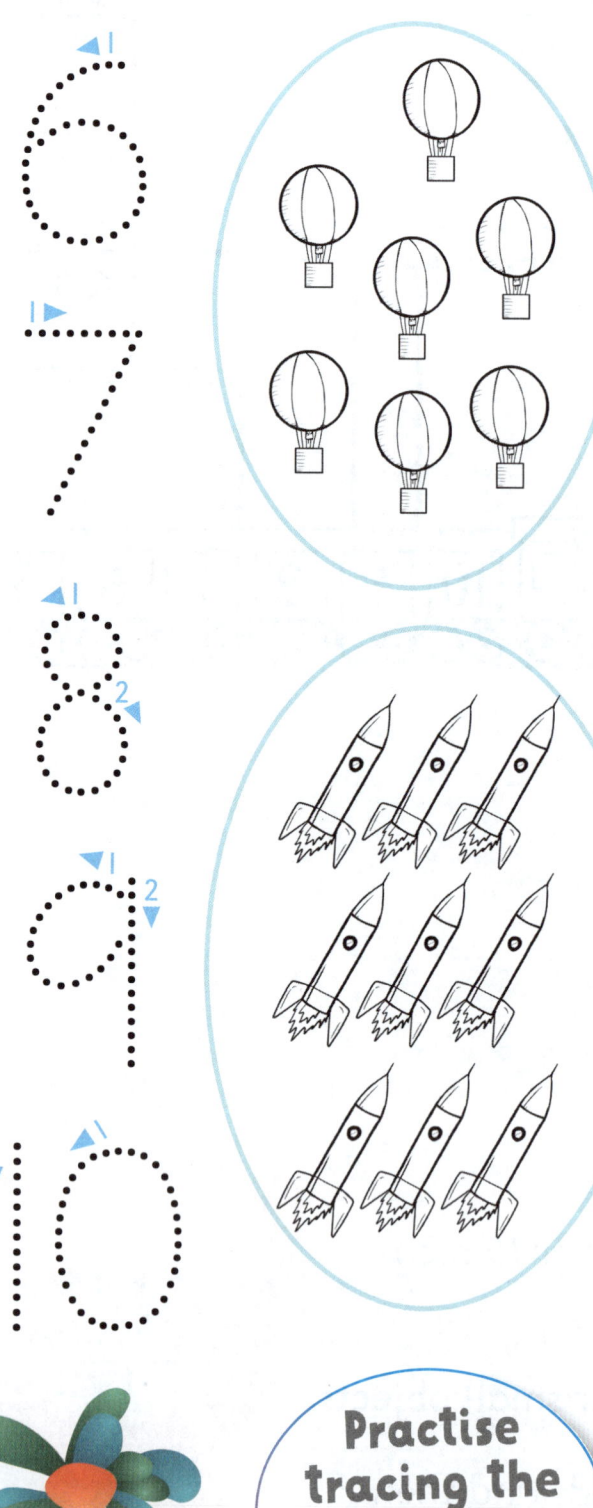

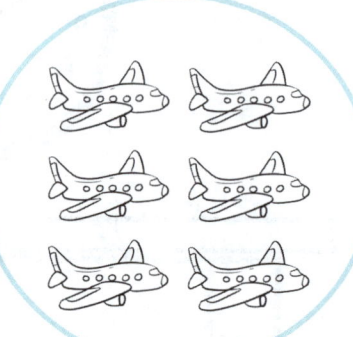

ten

six

nine

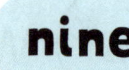

seven

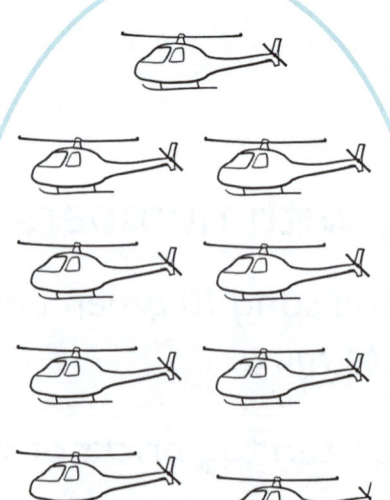

eight

Practise tracing the numbers too.

6
7
8
9
10

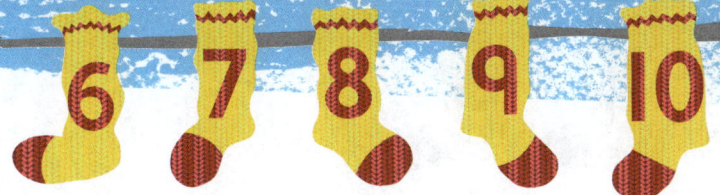

Count the apples on the trees.

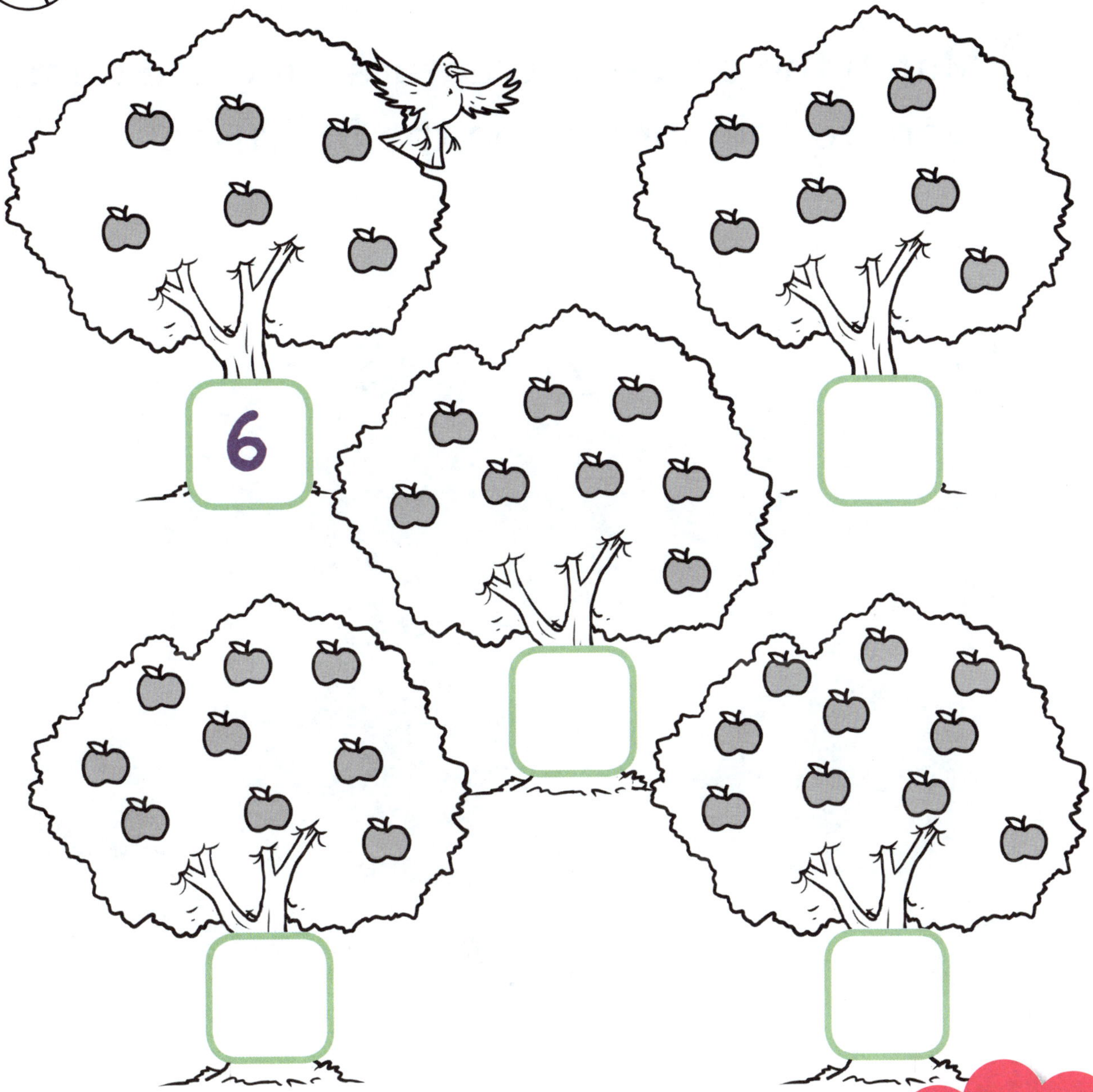

6

Write the number of apples on each tree in the boxes.

Now, draw a line to show the bird the way from tree 6 to tree 10 in order.

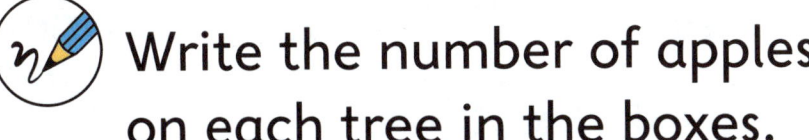

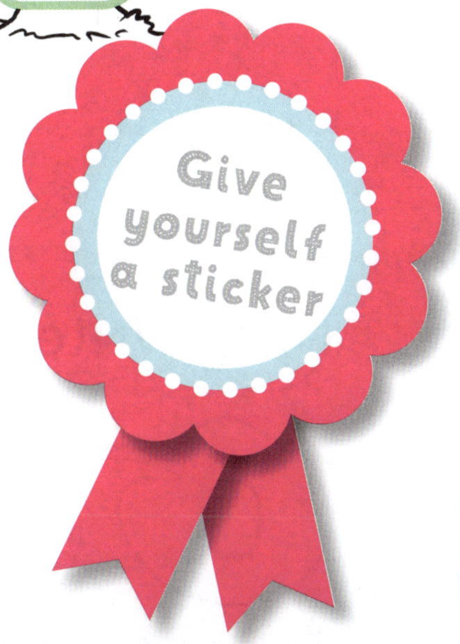

Give yourself a sticker

Now – track how you're doing on page 32!

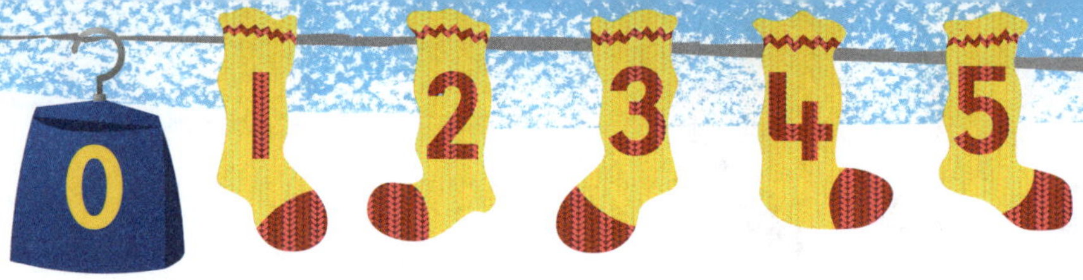

Numbers 0–10

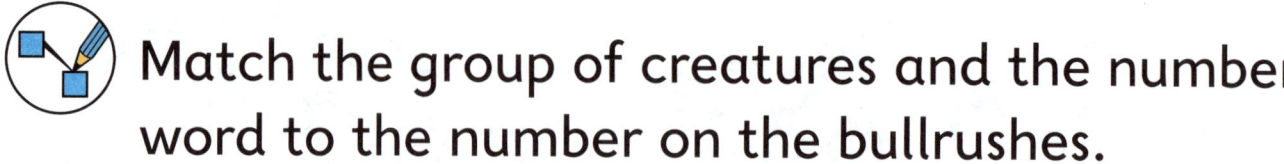 Match the group of creatures and the number word to the number on the bullrushes.

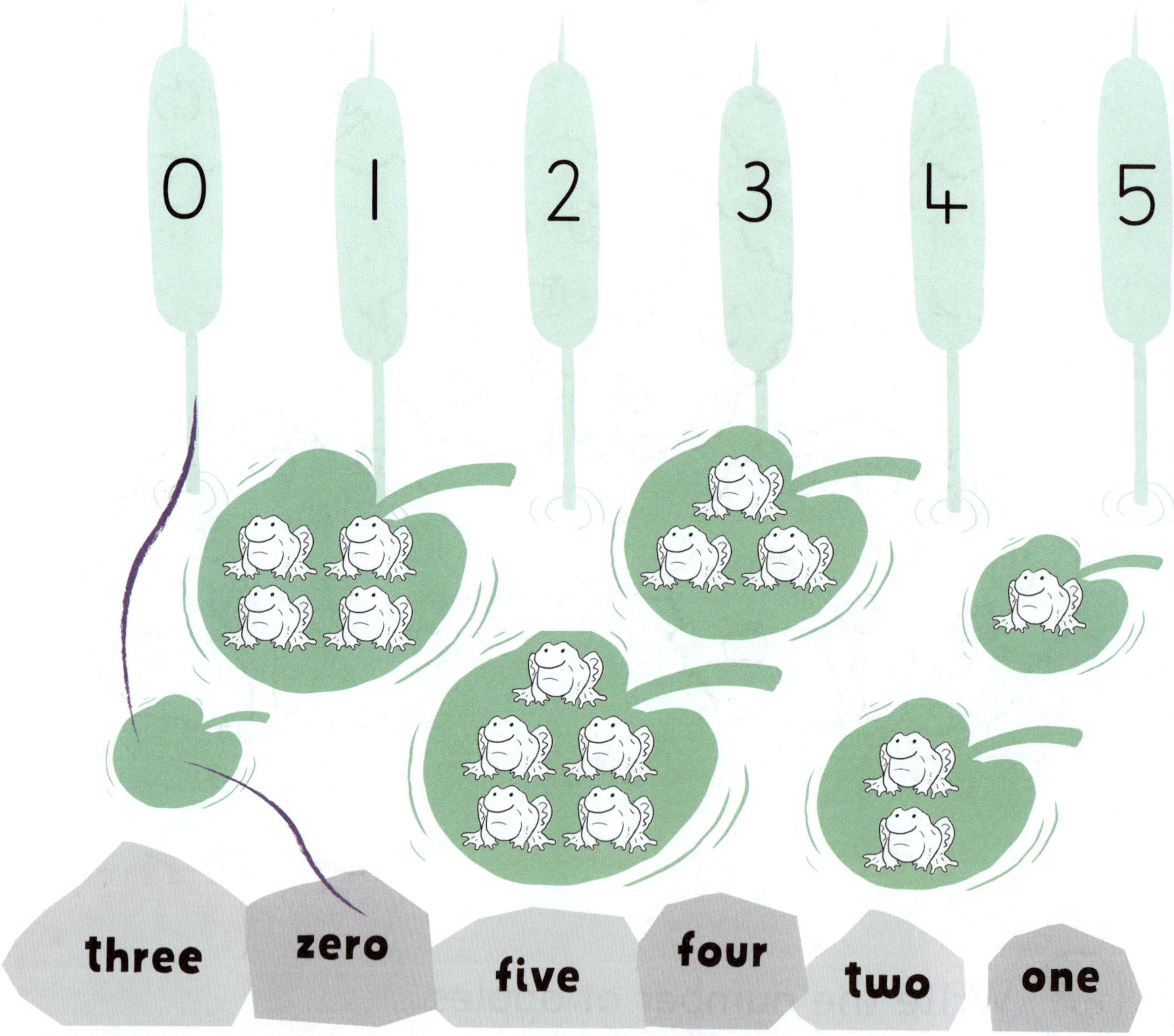

0   1   2   3   4   5

three   zero   five   four   two   one

✏️ Write the missing numbers on the lily pads.

0   3   5

nine   seven   ten   six   eight

Give yourself a sticker

Now – track how you're doing on page 32!

More numbers 0–10

 Write the numbers 0 to 5 in order.

_____

_____

_____

The penguin wants to eat the fish. Start at 0. Colour numbers in the correct order to make a path to 10.

	0	4	8	10	7
	1	2	6	3	2
4	7	3	0	5	8
1	5	4	9	10	
9	6	7	8	6	

 Write the numbers 6 to 10 in order.

_____

_____

 ⭐⭐ Circle the numbers, words and groups that come before **5**.

4   six

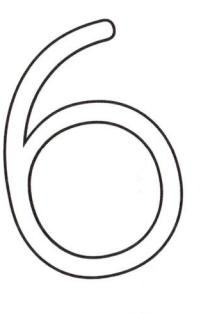 6   0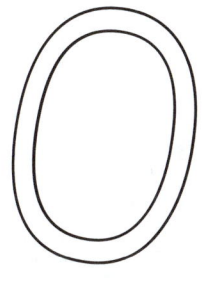

three

9   ten

zero

Say the numbers that come before or after 5.

Give yourself a sticker

 Now colour the numbers, words and groups that come after 5.

Now – track how you're doing on page 32!

Progress Chart

Colour in a face.

☺	I can do this well
☺	I can do this but need more practice
☹	I find this difficult

Page	I Can . . .	How did you do?
2–3	I can write number 0	☺ ☺ ☹
4–5	I can write number 1	☺ ☺ ☹
6–7	I can write number 2	☺ ☺ ☹
8–9	I can write number 3	☺ ☺ ☹
10–11	I can write number 4	☺ ☺ ☹
12–13	I can write number 5	☺ ☺ ☹
14–15	I know about numbers 0–5	☺ ☺ ☹
16–17	I can write number 6	☺ ☺ ☹
18–19	I can write number 7	☺ ☺ ☹
20–21	I can write number 8	☺ ☺ ☹
22–23	I can write number 9	☺ ☺ ☹
24–25	I can write number 10	☺ ☺ ☹
26–27	I know about numbers 6–10	☺ ☺ ☹
28–29	I can write and count 0–10	☺ ☺ ☹
30–31	I can write and count 0–10	☺ ☺ ☹

How did YOU do?